Practical Laboratory Skills Training Guides
High Performance Liquid Chromatography

Practical Laboratory Skills Training Guides

Coordinating Author: Elizabeth Prichard, *LGC, Teddington, UK*

Series titles:
Gas Chromatography
by Brian Stuart, *LGC, Teddington, UK*

High Performance Liquid Chromatography
by Win Fung Ho and Brian Stuart, *LGC, Teddington, UK*

Measurement of Mass
by Richard Lawn, *LGC, Teddington, UK*

Measurement of pH
by Richard Lawn, *LGC, Teddington, UK*

Measurement of Volume
by Richard Lawn, *LGC, Teddington, UK*

Also available:
Practical Laboratory Skills CD-ROMs

For further information please contact:
Sales and Customer Care, Royal Society of Chemistry, Thomas Graham House,
Science Park, Milton Road, Cambridge CB4 0WF
Telephone: +44 (0) 1223 432360
Fax: +44 (0) 1223 426017
Email: sales@rsc.org

Practical Laboratory Skills Training Guides
High Performance Liquid Chromatography

Win Fung Ho and Brian Stuart
LGC, Teddington, UK

Coordinating Author
Elizabeth Prichard
LGC, Teddington, UK

A catalogue record for this book is available from the British Library

ISBN 0-85404-483-3

© LGC (Teddington) Limited, 2003

Published for the LGC
by the Royal Society of Chemistry,
Thomas Graham House, Science Park, Milton Road, Cambridge CB4 0WF, UK
Registered Charity Number 207890

For further information see the RSC web site at www.rsc.org

Typeset by Land & Unwin (Data Sciences) Ltd, Bugbrooke, Northants
Printed by Athenaeum Press Ltd, Gateshead, Tyne and Wear

Preface

Production of this set of five Training Guides and CD-ROMs was supported under contract with the Department of Trade and Industry as part of the National Measurement System Valid Analytical Measurement (VAM) programme.

The guides were written by staff at LGC in collaboration with members of the SOCSA Analytical Network Group whose assistance is gratefully acknowledged. They include liquid and gas chromatography, the measurement of mass, volume and pH.

Training has formed an essential part of the VAM programme since its inception in 1988. Many training courses on topics aimed at improving the quality of measurements have been developed. However, in working with groups of analytical scientists it has become clear that the basic skills required in an analytical laboratory are not covered on courses or readily available in paper format.

These guides are aimed at filling this gap and are aimed at those working at the bench. For each topic they include a limited amount of theory to explain the essential features but the main emphasis is on what to do to ensure reliable results. They contain references to further reading for those who wish to study the topics in more depth.

To help laboratory managers assess the competence of the trainee there are a limited number of exercises suggested. The chromatography modules also have a trouble shooting section.

The CD-ROMs cover Practical Laboratory Skills and have links to websites where more information may be obtained.

Contents

High Performance Liquid Chromatography

1 Introduction

This booklet covers the basic practical aspects of high performance liquid chromatography (HPLC) and is aimed at the inexperienced analyst who may have no or very little knowledge of this technique. It includes basic tips, identifies key skills, arouses awareness and gives guidance on good practice of the basic aspects of HPLC. It provides information to help analysts in their understanding of the important issues to consider during analysis and to develop further skills.

2 Basic Theory

2.1 Overview

High performance liquid chromatography is one of the most widely used analytical techniques in industry. It is used to separate and analyse compounds through the mass-transfer of analytes between stationary and mobile phases.[1-3] The technique is employed in a broad range of activities, such as the analysis of foods, drugs and agrochemicals.

The technique of HPLC utilises a liquid mobile phase to separate the components of a mixture. The components themselves are first dissolved in a solvent and then forced to flow (*via* the mobile phase) through a column (stationary phase) under high pressure. The mixture is resolved into its components within the column and the amount of resolution is dependent upon the interaction between the solute components and the column stationary phase (immobile packing within the column) and liquid phase. The interaction of the solute with the mobile and stationary phases can be manipulated through different choices of both solvent and stationary phases. The individual units which form an HPLC chromatographic system are shown in Figure 1.

HPLC can be divided into two broad categories: normal phase and reversed phase. For normal phase, a polar stationary phase (usually silica) is used to retain

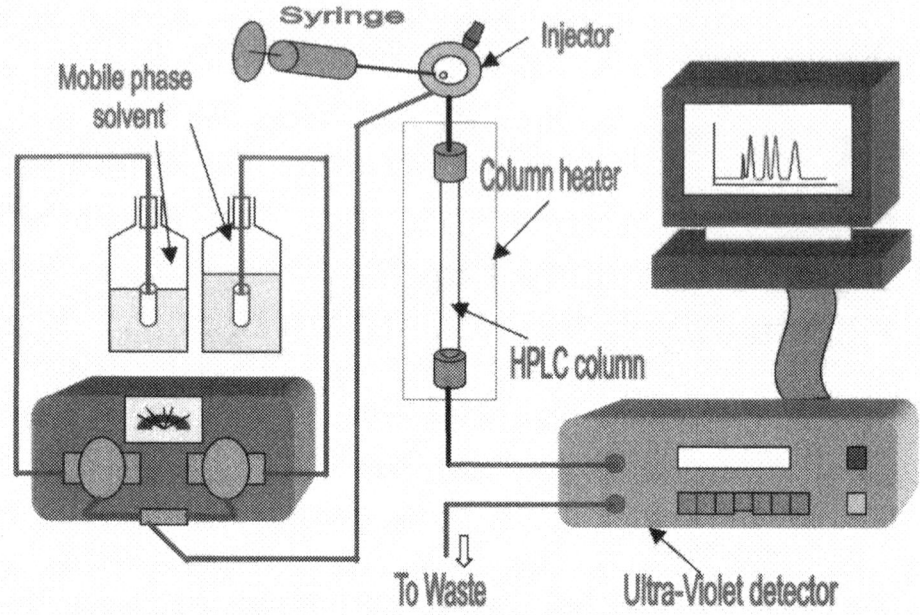

Figure 1 *Diagram of an HPLC chromatograph with ultra-violet detector*

analytes which are polar, whilst reversed phase separations are based upon forces between non-polar compounds and non-polar functional groups, which are bonded to the silica support. The majority of applications today are based on reversed phase separations. For a more in depth explanation of the principles and categories of HPLC, the analyst should consult references 1–3 and explore some of the websites listed in Section 9.

2.2 Key Parameters

This section describes some of the basic parameters that govern the effectiveness of a separation. The analyst should learn, by reading this section, which are the key separation parameters involved, their relative importance and how they can be calculated manually from a chromatographic trace. This is an important skill to master, despite the fact that with modern data handling software many of these parameters will be calculated automatically. For a more detailed explanation of those and other theoretical equations, it is advisable to consult references 1–4.

2.2.1 The Retention Factor (k)

The retention factor, k, is used to describe the migration rate of an analyte on a column and can be defined as shown in Equation 1 (see also Figure 2).

$$k = \frac{t_R - t_M}{t_M} \tag{1}$$

t_R retention time of a component
t_M dead time (time required for the mobile phase to pass through the column)

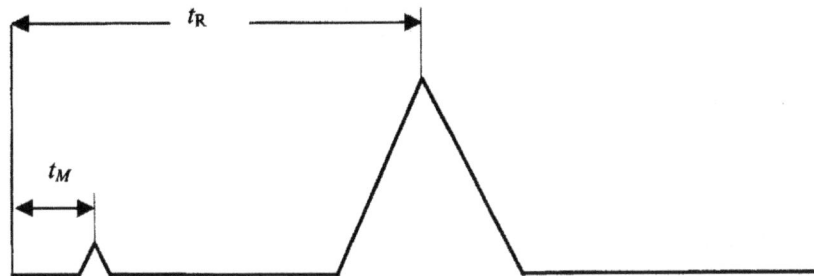

Figure 2 *Schematic of chromatogram showing parameters which determine the retention factor*

- Ideal separations are performed under conditions in which k is between 1 and 5.
- Components which take a long time to elute from the column compared to the mobile phase will have a large retention factor ($k > 20$).
- k Can be manipulated by varying the mobile and stationary phases.

2.2.2 The Selectivity Factor (α)

The selectivity factor, α, for two analytes within a column provides a measure of how well two components will separate on a column. The factor for a column with analytes A and B can be expressed as shown in Equation 2.

$$\alpha = \frac{(t_R)_B - t_M}{(t_R)_A - t_M} \qquad (2)$$

$(t_R)_B$ retention time of component B which is more strongly retained
$(t_R)_A$ retention time of component A which is less strongly held
t_M dead time

- When $\alpha = 1$, it is not possible to separate the two components using the given system.

2.3 Quantitative Measures of Column Efficiency

The plate model supposes that the chromatographic column contains a large number of theoretical plates. The sample equilibrates between the mobile and stationary phase in these plates, moving down the column by transfer from one plate to the next, as shown in Figure 3.

In reality, plates do not exist but the theory provides a model that serves as a way of measuring column efficiency. The terms plate height (H) and number of theoretical plates (N) are commonly used as quantitative measures to describe column efficiency. The relationship between the two is shown in Equation 3.

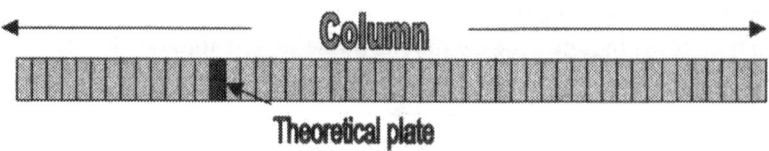

Figure 3 *Column as a series of theoretical plates*

$$N = L / H \qquad (3)$$

L   is the length of the column

An efficient new column producing sharp component peaks would be expected to exhibit a large number of theoretical plates (N). As the column ages and the peaks become broader, the efficiency of the column will drop accordingly. By measuring the number of theoretical plates for the column each time it is used, it is possible to monitor this deterioration over time. This can help the analyst determine whether the column is still 'fit-for-purpose', or whether it needs reconditioning or replacing. The value of N can be calculated from Equation 4 (see also Figure 4).

$$N = 5.54 \times (t_R / W_{1/2})^2 = 16 \times (t_R / W)^2 * \qquad (4)$$

t_R retention time
$W_{1/2}$ the width of the peak at 50% of the peak height
W   width of the peak at the baseline
5.54 and 16 are constants
* Chromatograms of samples often contain several peaks and it is not always possible to accurately measure the width of the peak at the baseline. An alternative solution is to measure the width of the peak at 50% of the peak height.

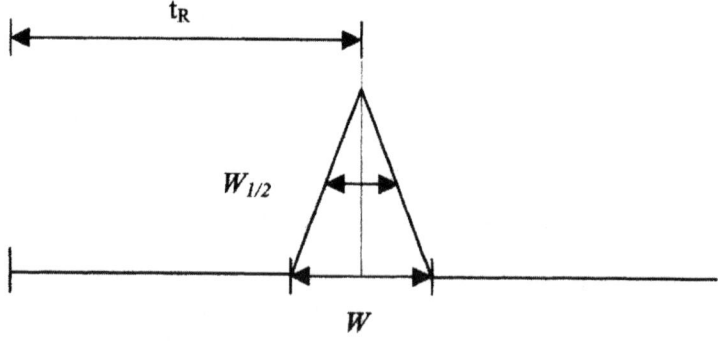

Figure 4 *Parameters that determine the number of theoretical plates*

2.4 Column Resolution

The resolution (R) of a column provides a quantitative measure of its ability to separate two components within a mixture. For a mixture with two components A and B, the resolution can be defined by Equation 5.

$$R = \frac{2[(t_R)_B - (t_R)_A]}{W_A + W_B} \qquad (5)$$

$(t_R)_B$ retention time of component B which is more strongly retained
$(t_R)_A$ retention time of component A which is less strongly held
W_B width of peak for component B at the base line
W_A width of peak for component A at the base line

- If R is less than 1, the components are overlapping.
- If R is equal to or greater than 1, this indicates good separation.

3 HPLC Components

3.1 The Mobile Phase

Selecting the correct composition and type of mobile phase is important because it is a variable that governs separation. However, choice is restricted because of the column used, *i.e.* the type of stationary phase employed. The main distinction is between reversed phase and normal phase chromatography. In normal phase systems, non-polar solvents such as hexane or isooctane are used whereas reversed phase requires polar solvents such as water, acetonitrile or methanol.

The choice of mobile phase is governed by the physical properties of the solvent. Factors to consider are: polarity, miscibility with other solvents, chemical inertness, UV cut-off wavelength and toxicity. The polarity index gives an indication of the ability of a solvent to elute a compound from the column.

Table 1 provides a summary of the typical solvents that are used for mobile phases and some of the important parameters that govern choice.

The HPLC system can be set up either for *isocratic* or *gradient* elution:

- Isocratic elution is where the mobile phase composition remains constant during the whole analysis.
- Gradient elution is where the mobile phase composition is steadily changed during the analysis, *e.g.* to obtain better resolution and/or decrease analysis time.

3.1.1 Practical Tips for Handling Mobile Phase

a) Consider the purity of solvents and only use HPLC grade materials:
   - impurities give rise to noisy baselines, *e.g.* UV detection at low wavelengths
   - de-mineralised water should be employed for mobile phases

Table 1 *Typical solvents for HPLC mobile phases*[5]

Solvent	Polarity index	UV cut-off /nm	Toxicity
Normal phase			
Hexane	0.1	210	Chronic neurotoxic
Isooctane	0.1	205	Low
Diethyl ether	2.8	218	Low
Dichloromethane	3.1	245	Chronic carcinogen
Isopropyl alcohol	3.9	205	Low
Reversed phase			
Water	10.2	200	None
Methanol	5.1	210	Mildly toxic
Acetonitrile	5.8	210	Toxic by inhalation
Tetrahydrofuran	4.0	280	Toxic by inhalation

b) Ensure that the mobile phase is free from dust; connect a stainless steel filter element to the end of the tube leading from the reservoir to the pump.

c) Remove dissolved air, because this can cause irregular pumping action and fluctuating signals from the detector, by performing one or more of the following:
   - degas the mobile phase with helium
   - place the mobile phase under vacuum
   - agitate the mobile phase in an ultrasonic bath

d) When mixing solvents to form mobile phases:
   - the analyst must understand the terminology which is used to describe the constituents of the mobile phase – a common expression is for example, 75/25 v/v methanol/water which indicates a volume measurement, *e.g.* 75 mL of methanol + 25 mL of water)
   - most systems have a facility to mix the solvents from different reservoir bottles – the analyst only has to set the percentages (*e.g.* 75% methanol, 25% water)
   - if a mobile phase of mixed composition is to be prepared manually, the volume of each solvent should be measured separately before they are mixed together

e) Be aware that volatile components in a mobile phase of mixed composition may evaporate. This can be minimised by:
   - keeping the solution cool during the degassing procedure
   - keeping the reservoir bottle stoppered at all times

f) Ensure that the sample to be analysed is soluble in the mobile phase.

g) When using UV detectors you must consider the UV absorption of the mobile phase. This is indicated by the UV cut-off value. For example, tetrahydrofuran has a UV cut-off of 280 nm (Table 1), therefore it cannot be used for analysis of samples for pyridine as the peak maximum for pyridine is ≈ 260 nm (Figure 14).

h) Ensure that the mobile phase does not react with the stationary phase. Buffers and pH modifiers may contain, for example, ammonia which can replace one of the N–R groups of an amide on an aminopropyl stationary phase.

i) It is also important to monitor the levels of the mobile phases and ensure that they are constantly topped up and therefore the system is never allowed to run dry.

j) Whenever there is a change in the mobile phase, ensure that the labels on the bottles are also changed and labelled correctly with the following information:

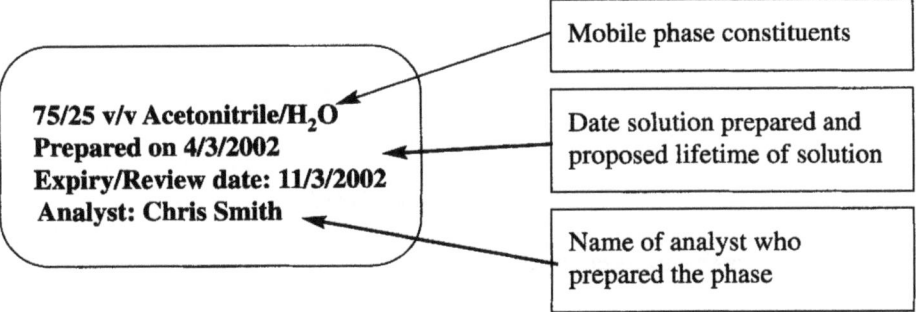

Mobile phase constituents

75/25 v/v Acetonitrile/H_2O
Prepared on 4/3/2002
Expiry/Review date: 11/3/2002
Analyst: Chris Smith

Date solution prepared and proposed lifetime of solution

Name of analyst who prepared the phase

3.2 The Pump

The main purpose of the pump in HPLC is to pass a constant flow of mobile phase through the chromatographic column. There are two main types of pump used in HPLC, incorporating either a syringe or reciprocating piston element in their design.

3.2.1 Syringe Pump

Syringe-type pumps are attractive because they operate pulse free. However, the total volume of mobile phase that the pump can deliver is limited by the capacity of the syringe. The pumps are expensive in comparison with the more commonly used reciprocating piston type.

3.2.2 Reciprocating Piston

A schematic of this type of pump is shown in Figure 5. A rotating cardioid cam causes the two pistons alternatively to draw mobile phase from the reservoir and then to force the liquid in the direction of the column.

The check valves are opened or shut in synchrony with the piston movement to ensure that flow of mobile phase only occurs in one direction. The volume delivered

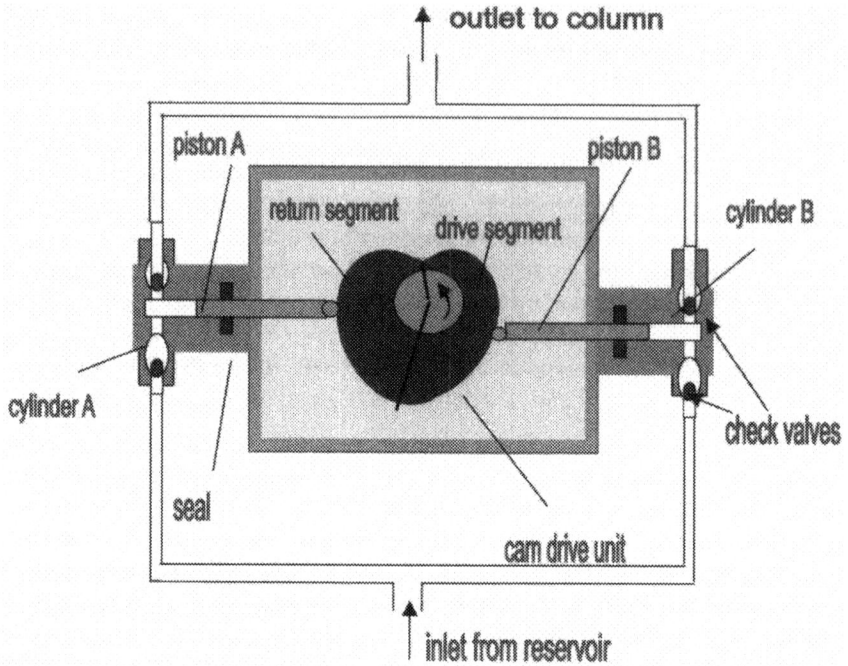

Figure 5 *Dual piston reciprocating pump*

by each piston per cam revolution is typically 0.1 cm^3. The design is relatively simple and has the advantage that a very large reservoir of mobile phase can be accommodated to allow almost continuous use of the equipment. Flow rate can be altered simply by changing the speed of the motor driving the cam. This type of pump has the disadvantage that it causes pulses in the flow of mobile phase.

3.2.3 Other Considerations Concerning the Pump

a) The pump must be able to deliver the mobile phase at high pressures in order to overcome the flow resistance associated with HPLC columns. High working pressures are particularly important when microbore columns, smaller particle size packings, high flow rates and viscous mobile phases are used in the analysis.

b) The components of the pump must be resistant to corrosive chemicals and solvents. Acids, bases and aggressive organic solvents are commonly used in mobile phase formulations so the analyst should be aware that these chemicals can cause damage to the equipment.

c) Flow rate should be easy to set with a possibility of settings between ~0.1 and 10 mL min^{-1}.

d) Pumps should be robust and should be able to function routinely with only a minimum requirement for maintenance and servicing.

e) Flow should be pulse-free and stable. Pulses, fluctuations in the flow of mobile phase, are always undesirable and can lead to poor precision for component retention times. If the detector attached to the column is 'flow sensitive' the chromatogram baselines may exhibit peaks and troughs which mirror the changing flow rate. This increased noise leads to raised detection limits for the sample components.

f) The pump should be a 'constant flow' device. Constant pressure pumps do not give satisfactory performance because the flow rate will vary as the resistance to flow offered by the column changes, *e.g.* when internal sinters are blocked by particulates and columns become overloaded with sample residues.

3.2.4 Tips for Pump Use[6]

● After use, always flush water, buffered solutions, acids, bases and aggressive organic solvents out of the pump using an appropriate solvent. This prevents damage to seals, valves and pump heads.

● Always degas mobile phases to prevent bubble formation in the pump.

● Check the pump is delivering the mobile phase at the desired flow rate – measure the actual flow rate by measuring the time it takes for a fixed volume of mobile phase to exit the column.

● Always ensure, when using gradient elution, that the components of the mobile phase are miscible for *all* compositions employed in the method. Buffer salts have a tendency to precipitate from solution as the content of the organic solvent in the mobile phase is increased.

● Prime the pump before use to confirm that there are no bubbles in the cylinders.

● If the mobile phase is to be changed, ensure that the new phase is miscible with the previous phase; if not, employ an intermediate solvent to thoroughly flush out the old mobile phase. It is also advisable to purge the system to remove any air bubbles that may have been introduced during the change.

3.2.5 Other Considerations and Rules for Using HPLC Pumps

a) Never operate a pump without a solvent reservoir filter on the inlet tubing. Unfiltered solvent may block pumps and/or shorten seal lifetime which could adversely affect the check valve.

b) Always check the pump for leaks before and during analysis, this is especially important when the system will be unattended for long periods of time.

3.3 The Injector

The HPLC injector, unlike its GC equivalent, is usually very simple in design. Figure 6 shows a schematic of a typical loop injector. The injector is set up in the 'load' position and the sample is injected, using a blunt-tipped syringe, through a needle port into an open loop. Sufficient volume of sample is injected to fill the loop and any excess liquid is allowed to exit the loop *via* the vents. The injector is then switched to the 'inject' position where the loop ends are connected to the pump

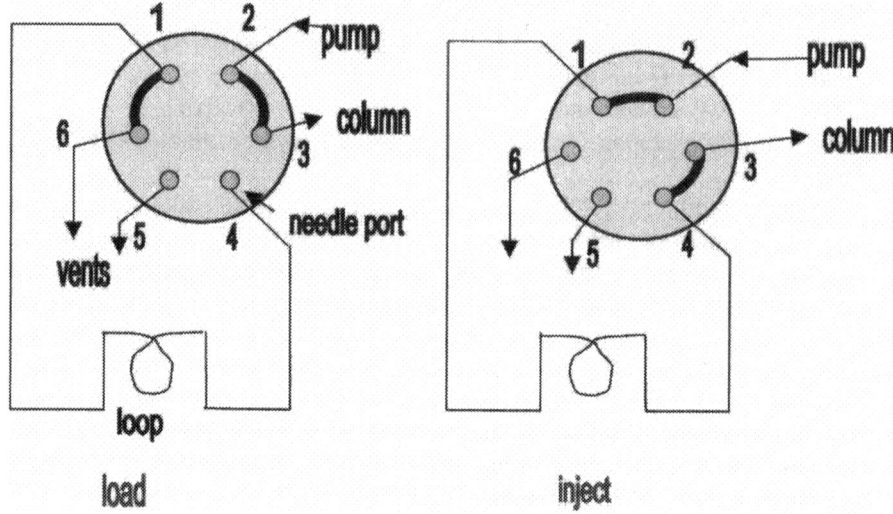

Figure 6 *Schematic of a loop injector (Rheodyne 7125)*

and column. The sample slug (occupying the loop) is then carried by the flow of mobile phase to the top of the column. This design allows the injector loop to be loaded at ambient pressure so the problems associated with high column back pressures and leaking injectors and septa are eliminated. Because the loop volume is fixed the injection process is highly reproducible allowing the HPLC to be used for quantitative analysis. Loops are interchangeable allowing the injection volume to be selected according to the analysis required.

3.3.1 Tips for Use of Injector

- Never use syringes with sharp tips because these will damage the injector.
- Check the injector for leaks prior to the analysis.
- Never allow the loop to siphon during the loading sequence; this can allow air to enter the loop. Long lengths of vent tubing positioned vertically can cause siphoning.
- Ensure that the column is not overloaded with sample – select the loop volume that is appropriate for the size of the HPLC column used. The load for analytical columns can be in the range 0.5 to 100 µL but this has to be determined for the particular column being used.

3.4 The Column

3.4.1 Introduction

The column is the most important chromatographic component and it is crucial in determining the performance and resolution of the whole HPLC system. The choice of column is governed firstly by the type of chromatography used, *i.e.* reversed phase or normal phase chromatography. Figure 7 summarises the types of stationary phases and their relative usage from a survey conducted by *LCGC* magazine (1998) and Table 2 shows some typical applications.

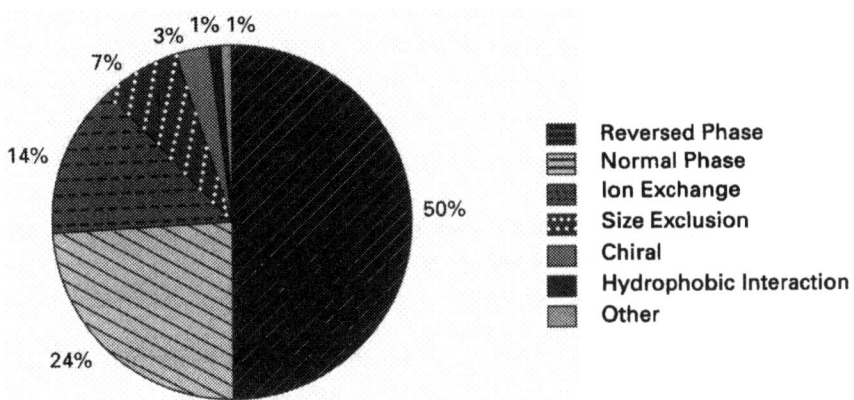

Figure 7 *HPLC mode and stationary phase usage*

Table 2 *HPLC stationary phases and their typical applications*

Stationary phase	Type	Application	Mobile phase	Typical analytes
Silica		Normal phase	Hexane, alcohols	Pesticides and natural products
Octadecylsilyl (ODS)	C18 hydrocarbon chain	Reversed phase	Water, methanol, acetonitrile, buffers (pH 2–8)	Peptides and amino acids
C8	C8 hydrocarbon chain	Reversed phase	See ODS	Drugs and pharmaceuticals
Cyanopropyl (CN)	Cyanopropyl group bonded to silica support	Both normal and reversed phase	Reversed phase – water, alcohol Normal phase – hexane, ether	Foods and fatty acids
Aminopropyl (NH$_2$)	Aminopropyl group bonded to silica support	Both normal and reversed phase	Reversed phase – water, alcohol Normal phase – hexane, ether	Surfactants
Pirkle	Phenylglycine enantiomer bonded to a silica support	Chiral ionic separations are more effective but less robust	Hexane, modifiers	Pesticides and herbicides

For further details, the analyst should refer to their column manufacturers/ suppliers. *NB:* Many manufacturers now have their own website. These contain useful information about the types of column that are available along with suitable applications.

3.4.2 Reversed Phase Columns

Why are reversed phased columns so popular?

- They are versatile and their stationary phase chemistry meets the needs of a wide range of samples.
- The stationary phase material can be manufactured to high quality and is reasonably stable under normal operating conditions.
- The elution order of the compounds being separated can be predicted easily and is based upon their hydrophobicity.
- Mobile phases are simple mixtures of aqueous solvents. Water, which is usually the predominant component, is inexpensive, safe and plentiful.

3.4.3 Types of Columns

There are many types of reversed phase columns, which are distinguished by the functional group that is bonded to the silica support. Figure 8 shows some of the common types of reversed phase columns. For a fuller understanding of the reasons behind the development of reversed phase packings, see reference 7.

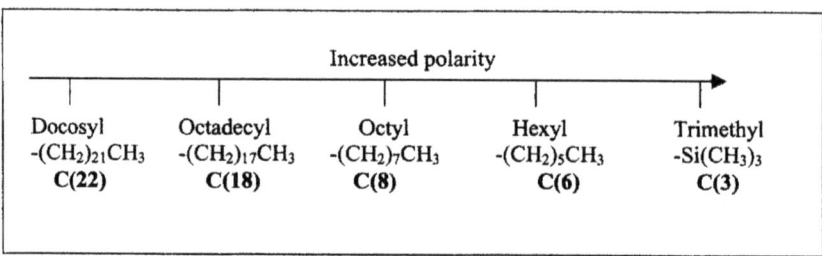

Figure 8 *Typical reversed phase columns and their relative polarity*

Octadecylsilane (ODS) columns are the most commonly used and are usually the preferred choice to begin analyses of unknown non-polar compounds (see Figure 9). This is a very non-polar phase and polar compounds will therefore be eluted faster than their non-polar counterparts. A C8 column might therefore be more useful than the C18 analogue if the compounds to be analysed are polar.

Column selection is governed by a number of factors, these include:

- reviewing data from previous work on similar compounds
- using the knowledge of the chemical structure of the compounds to be separated

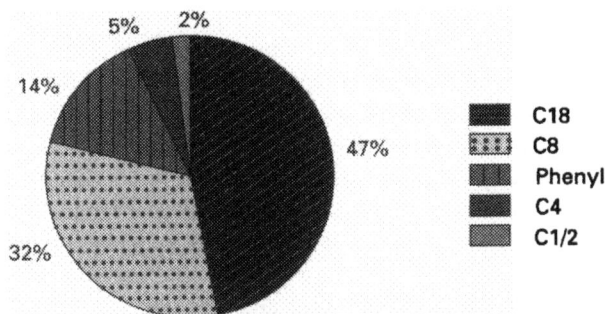

Figure 9 *Stationary phases used in reversed phase chromatography*

- consulting the literature, *e.g.* look at the information provided in the manufacturer's catalogue
- analysis of a trial sample to test if the selected column performs satisfactorily

3.4.4 Column Characteristics

Once you have decided which phase to use, it is important to consider other column characteristics which may affect the analysis. Even if you are not required to choose a column, you should at least know and understand some of the fundamental column properties that control separation.

3.4.4.1 Column length and internal diameter
Some typical values for the column parameters are shown in Table 2.

Table 2 *Column parameters*

Column type	Internal diameter/mm	Injection volume	Flowrate/$\mu L\ min^{-1}$
Conventional	4.6	5–20 μL	500–2000
Narrow bore	2	1–5 μL	200
Microbore	1	0.5–2 μL	50
Capillary	0.5	0.1 μL	5
Nanoflow	0.05	nL	0.1

- Columns are typically 15 cm long with an internal diameter of 4.6 mm but alternative lengths and internal diameters are available.
- Column dimensions determine the flow and sample injection volume used and greatly influences the length of time for an analysis.
- The current trend is to use shorter lengths (3 or 5 cm) and smaller internal diameters (1 or 2 mm) for quicker analysis.
- Capillary columns (with diameters ranging from 180 to 300 μm) are becoming more popular due to quicker analysis and savings because less solvent is consumed.

3.4.4.2 Column support material

Most support materials (to which the functional group is bonded) used in HPLC column manufacture are made from silica. Manufacturers of HPLC columns offer a wide range of porous silicas, which are characterised by pore size, particle size and surface area. The particles can be either spherical or irregular in shape. The latter is less efficient but usually cheaper and is used mainly for large-scale preparative applications. The name of the column is usually derived from the type of silica used for packing and examples of commercially available materials include:

- spherical silica – Nucleosil, Spherisorb and Hypersil
- irregular silica – LiChrosorb, Partisil and Sorbsil

3.4.4.3 Choosing particle size

The particle size of spherical packing materials range from 3 μm to 20 μm in diameter. The size affects column efficiency and back pressure. The efficiency of the column decreases as particle size increases so there is poorer peak separation. However, the flow resistance of the column also decreases with particle size which means that the back pressure decreases. The most popular particle size for analytical columns is 5 μm.*

3.4.4.4 Pore size and surface area of support materials

Packing materials with a large pore size will have a small surface area whereas those with a small pore size have a large surface area. For most applications, analysts should select a column material with a small pore size (<10 nm) because the increase in surface area improves the capacity of the column. In practice column materials have a pore size distribution that spans about an order of magnitude. However when analysing samples containing high molecular weight materials, *e.g.* proteins, a column with a large pore size will be required (>>10 nm)

3.4.4.5 Column specification – a summary

Columns are typically described with the details in Figure 10. These are the specifications that that are required when purchasing a HPLC column from a manufacturer.

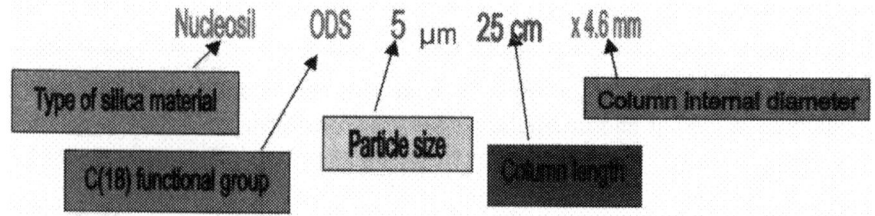

Figure 10 *Column specification*

* Information from www.waters.com

3.4.5 Guard Columns

Guard columns are employed as a protective factor and are installed between the injector and the main HPLC column. Essentially, they are designed to filter and remove unwanted particles that may clog up the main column. Use of a guard column often prolongs the life of the main column. The total internal volume of the guard column should be small to minimise peak broadening.

3.4.6 Column Care – Practical Tips

3.4.6.1 Monitoring column performance
A new column is usually required when the column has either been damaged and/or degraded with time. Usual indicators of inefficient performance include:

- consistently poor peak shapes and noisy baselines on the chromatogram
- a significant decrease in retention times and poor peak separation

This can be confirmed by measuring the column efficiency (see Section 2.3), *i.e.* using a test sample mixture to determine the number of theoretical plates. This value can be compared with the original column efficiency when the column was purchased to determine if there has been a notable deterioration in performance. Many columns come with a certificate of analysis and a test sample chromatogram showing the original column's performance. Performance in the short term may be improved by carrying out column cleaning procedures recommended by the manufacturer.

If it is often necessary to change the column for a different method. The steps involved in the change over are shown in Figure 11.

3.4.6.2 Optimise column performance
To improve column performance, the following points must be considered:

a) Avoid extra connections between the injector and column, and between the column and detector. Use only one piece of capillary tubing between the injector and column, and between the column and detector.

b) pH stability is important because the operating conditions must be kept within the ranges specified by the manufacturer. Any deviation outside the limits will affect the column and undoubtedly decrease its life. Most systems employ a buffer solution to control and maintain the pH stability of the mobile phase. For example, if the pH is not controlled when using ODS columns, the following might occur:
   - at low pH (<2.5), the siloxane bonds can be destroyed by hydrolysis
   - at high pH (>8.5), silica dissolution can occur

c) Maintain the column at constant temperature (typically 40 °C) by housing it in an oven. Variation in temperature can cause inconsistencies such as problems in the identification of components from the chromatogram due to changing retention times. It is also necessary to check that the temperature reading

Flush out the column with the appropriate storage solvent (e.g. methanol or propan-2-ol for an ODS column) – have regard to miscibility of solvents when changing from the method mobile phase to the storage solvent. Allow sufficient time for the storage solution to replace the mobile phase in the column. A typical HPLC system with a 4.6 mm × 250 mm column and flow rate 1 mL min^{-1} needs approximately 10 minutes to change the solvent.

Switch off pump or put on standby mode.

Remove the column. The column ends must be immediately sealed with the appropriate caps (which should be kept safely in a drawer) to prevent the column from drying out and producing voids that will deteriorate its performance.

Store column in a safe place, e.g. in its original transportation container. Keep a record of when it was removed from the system.

Check the new column is being fitted the correct way. An arrow is marked on the side of the column to indicate the direction of flow of the mobile phase.

Install the new column and check that the connections are consistent with those already in place, if not, use appropriate connectors. Connections need to be clean and aligned correctly to get a good seal. Once fitting has been completed, ensure that connections are adequately tightened so as to prevent leakage. Beware of over-tightening components because this may cause unnecessary stress to joints that may result in leaks within the system.

Flush the new column with the mobile phase intended for its use. Check for miscibility problems between the column storage solvent and the mobile phase. Typically pump through ten column volumes of mobile phase before use. Allow the new column to settle to the required operating temperature within the oven, prior to use. NB: Flow rates are dependent on the diameter of the column. Typical flow rate for a 4.6 mm column is 1 mL min^{-1}.

Figure 11 *Steps to follow when changing a column*

corresponds to the actual temperature of the oven. Using a column oven is beneficial because:

- reproducibility of analysis is improved
- consistent retention times are promoted
- drift due to laboratory temperature and seasonal fluctuations are eliminated

d) Testing – the performance of the column should be monitored on a regular basis. This is performed by running a standard test mixture through the HPLC system. *NB*: Manufacturer's columns are often supplied with a chromatogram of a standard test mixture. The column should be tested on receipt with the test mixture to identify if damage has occurred in transit. The test mixture should be run on a regular basis when the column is in use. Data such as retention times and peak areas of individual components can be monitored to determine if they have changed significantly with time. Any significant change means that there is a problem with the column or HPLC system and remedial action is required. The chromatograms of the test mixture obtained should be archived to facilitate easy checking of the column's performance over time.

e) Cleaning – column cleaning is recommended and details of solvents to use are usually included in manufacturers' literature. The appearance of unexpected peaks and/or drifting baselines on your chromatogram is usually a good indicator of a dirty column. Most analysts will use components of the mobile phase to flush columns in an attempt to remove impurities that may have built up with time. For example, if using ODS columns, it may be possible to wash with a series of solvents such as water, tetrahydrofuran and acetonitrile. Column frits, which are porous elements located at the ends of the column (their purpose is to retain the column packing) can sometimes get blocked and may require cleaning and/or replacement. Additional guidance and tips on column washing can be found in reference 8.

3.4.7 Column Lifetime

Columns are (often) expensive and must therefore be treated with care. It is essential to use them in accordance with manufacturers' recommendations. For additional details on the care and maintenance of modern columns see reference 9. Tips of how to maximise the life of a column are indicated in Figure 12.

3.5 The Detector

3.5.1 Ultra-violet Detectors (UV)

UV detectors are the most commonly used detectors because they can be used to analyse a range of organic compounds and are relatively simple to use. A cross-section of a UV flow cell is shown in Figure 13.

When UV-absorbing sample components flow from the column into the light path of the flow cell, the UV radiation is absorbed and a reduced signal is observed at the detector. Because each component passes through the flow cell as a band of solute, the detector records a normal gaussian shaped peak. The ability of a compound to absorb UV (or visible) radiation is dependent on the chromophores in the molecular structure.

The UV detector is used extensively in the pharmaceutical, environmental, food and agrochemical sectors because the analytes involved usually contain chromophoric groups such as, thiol, carbonyl, alcohol, amine, ethylenic, acetylenic, nitrite, sulfone, iodo, phenyl and heterocyclic.

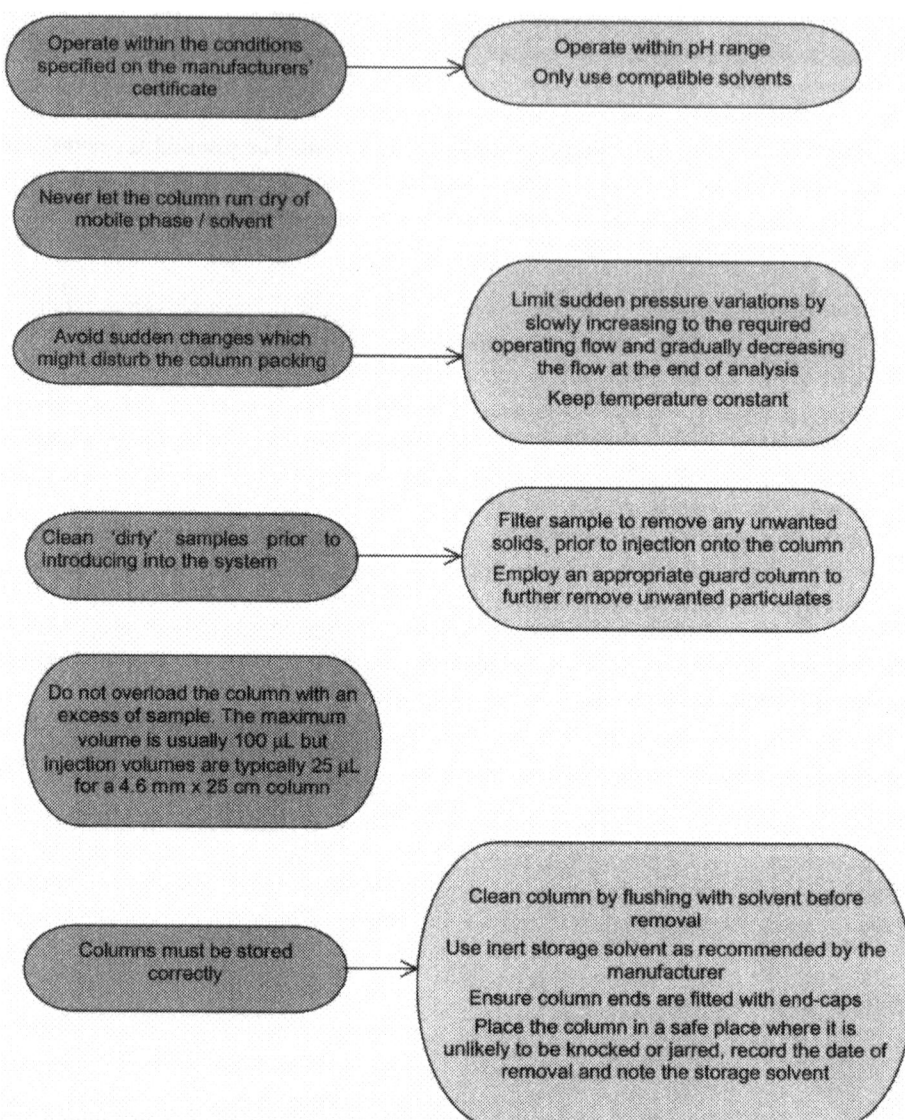

Figure 12 *Tips on how to maintain and extend column life*

The UV spectrum of pyridine in Figure 14 shows that the compound absorbs strongly in the UV with a maximum absorption near 260 nm.

3.5.1.1 Types of UV detector

Fixed wavelength measures at one wavelength (typically 254 nm).

Variable wavelength measures at one wavelength at a time but can detect over a wide range, *e.g.* 190 to 400 nm. Such detectors offer the best sensitivity for any absorptive component by allowing the user to select an appropriate wavelength at

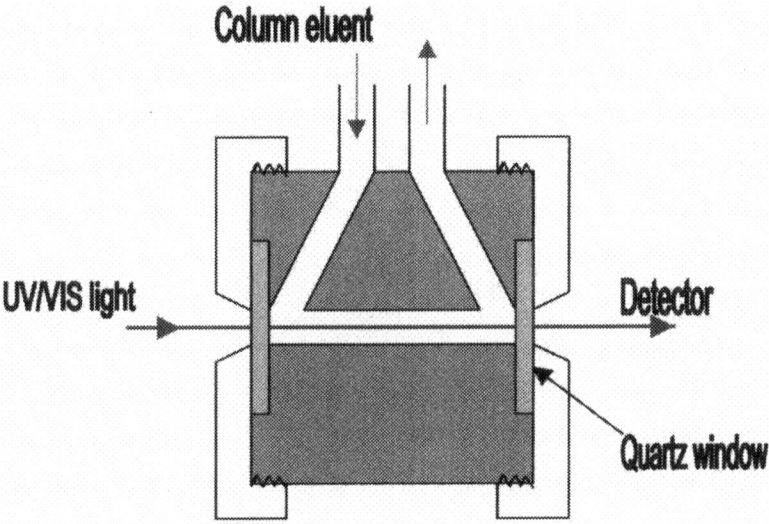

Figure 13 *U-Shaped geometry flow cell*

which to monitor. Individual analytes may have high absorptivity at different wavelengths and thus, single wavelength detection would reduce the system's sensitivity.

Diode array measures a spectrum of wavelengths simultaneously and allows the user to perform spectroscopic scanning to determine precise absorbance readings at a variety of wavelengths while the peak is passing through the flow cell.

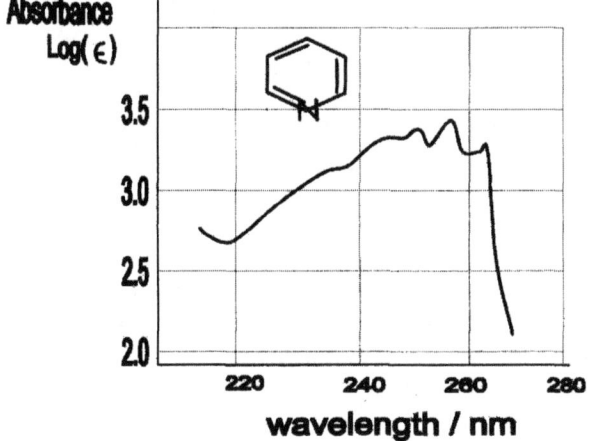

Figure 14 *The UV spectrum of pyridine*

3.5.1.2 Choice of wavelength for UV detectors

If working in accordance with established methods, wavelength settings are specified. Otherwise, they can be determined from the literature or by measuring the absorption spectrum of the analytes. For example, aromatic rings show strong absorption near 254 nm and this is a popular choice of wavelength for detecting compounds of this nature.

3.5.1.3 Maintenance awareness

- UV lamps must be replaced before they get too old and become inefficient.
- Monitoring of lamp performance is recommended and its intensity should be checked against manufacturer's specifications.

3.5.2 Refractive Index Detector (RI)

A schematic of the RI detector is shown in Figure 15.

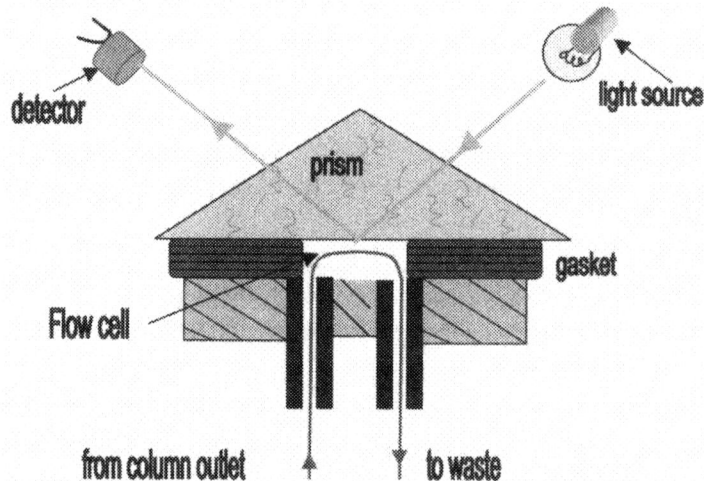

Figure 15 *Schematic of the refractive index detector*

In this schematic the magnitude of the detector signal is a function of the refractive index of the liquid in the flow cell, *i.e.* the mobile phase. When the sample components enter the flow cell the refractive index of the liquid is altered and the detector response changes accordingly.

In the differential RI detector a reference flow cell is located next to the sample flow cell. This is filled with mobile phase and is illuminated by the same light source. Here the detector output corresponds to the difference in signal arising from the sample and reference beams. Detectability is increased using this design and the detector is less sensitive to changes in source intensity and detector temperature.

3.5.3 Other Detection Systems

There are several types of detection systems and a summary of these detectors and their typical applications is shown in Table 4.

Table 4 *HPLC detectors*

Detector	Application	Comments	Cost
Fluorescence	Fluorescent compounds	High sensitivity and very specific	Moderate
Chemiluminescence	Chemiluminescent reactants or activators	Very sensitive and highly specific	Cheap (often home made)
Electrochemical	Acids and bases	High sensitivity and very specific	Cheap
Mass spectroscopy	Universal	Reasonably sensitive but complex to use	Expensive

A more comprehensive review of detectors can be obtained from references 1–4.

4 HPLC System Parameters and System Suitability Checks

4.1 System Parameters

Apart from the column and mobile phase there are several other significant parameters within the HPLC system which, if altered, can affect the separation procedure during analysis.

4.1.1 Flow Rates

The flow rate is invariably specified in established methods. It is important to check that the correct flow rate has been selected (in accordance with the written procedure) and that the value displayed agrees with the volume being delivered. Changes in the flow rate from the set values are usually identified from the chromatogram. For example, inconsistent peak retention times and drifting baselines are indicators of possible flow problems.

   Flow rate fluctuation may be due to:

- problems with the pump
- blockages within the system
- leaks within the system

Check that flow rate(s) are consistent with the programmed values by measuring the amount of mobile phase that comes out from the waste (into a measuring cylinder), over a specified period of time.

4.1.2 Temperature Changes

The effect of temperature on the chromatographic processes in HPLC is quite small in comparison to that found with GC. Although there is a performance benefit associated with an increase (above ambient) in the column temperature, in practice most analytical methods specify temperatures around 25–40 °C. Column thermostats have become an essential feature in modern instrumentation, because it is important to control the temperature of the column. Large fluctuations in the temperature will alter peak retention times, which may result in mis-identification of peaks. Therefore, the temperature of the column should be kept within ±2 °C.

Check oven temperatures for consistency by inserting a calibrated temperature probe/thermometer into the oven and monitor over a specified period of time.

4.1.3 Detector Sensitivity and Wavelength (When Using UV Detectors)

Changes in sensitivity and wavelength will critically affect the size of peaks on the chromatogram. Any deviations from set values will therefore alter the results and hence, it is important to know the exact operating values and that these have been correctly set, prior to running samples.

Before beginning an analysis check, by observation, that the settings are correct.

4.1.4 Integration Parameters

Computer software is invariably used to determine the peak areas (or peak height) in the chromatogram. The software works on integration parameters that can be set by the analyst. Such parameters include:

- peak width
- slope sensitivity
- threshold
- baseline setting (*e.g.* maintaining a horizontal baseline from the point where the signal was first recorded may not be appropriate)

The analyst must check that the integration parameters are appropriate for the sample chromatogram. For example, if the peak width has been set too narrow then noise might be interpreted as peaks whereas if it is set too wide then the software will not correctly integrate two peaks that are adjacent to each other in the chromatogram.

4.2 System Suitability Checks

System suitability testing should be conducted regularly and is essential when a new analysis is to be performed or changes have been made such as repair or replacement of parts.

The parameters which should be monitored are listed; they may be calculated from chromatograms using the equations listed in Section 2 of this guide or *via* HPLC computer software.

- Column performance:
  - the number of theoretical plates (N)
  - the resolution of a column
- Repeatability of injection:
  - repeat injections of a standard solution should provide a coefficient of variation which is no greater than ±1.0% of the mean peak area

4.3 Equipment Qualification Tests (EQ)

EQ is a documented process which provides evidence that an instrument is fit for purpose and kept in a state of maintenance and calibration consistent with its use. Detailed information on the EQ procedure and suggestions of what tests should be conducted are described in reference 10.

5 HPLC Test

5.1 Introduction

The following series of experiments will provide the analyst with an example of how changes to the parameters of the HPLC can affect the results.

Use an HPLC chromatograph system with a UV/VIS detector and a reversed phase column, *e.g.* an ODS column (25 cm × 4.6 mm i.d.).

Suggested initial operating conditions are:

Mobile phase	65% acetonitrile/35% phosphate buffer pH 3.2
UV wavelength	254 nm
Temperature	ambient
Injection volume	1 μL
Flow rate	1 mL min⁻¹

Prepare a mixed solution containing each of the following (in order of elution):

uracil, pyridine, phenol, dimethylaniline, 4-butyl benzoic acid, toluene.

The suggested volume is 100 mL as virtually all laboratories keep 100 mL volumetric flasks, however 10 mL would be sufficient.

a) Prepare a set of 1 mg L^{-1} solutions by weighing out 100 mg ± 5 mg of each of the compounds, transferring to separate 100 mL labelled volumetric flask and making up to the mark with acetonitrile. Stopper and shake gently to mix.
b) Prepare a mixed stock solution by pipetting, with an automatic pipette, 5 ml of each solution (2 mL in the case of uracil) into a 100 mL labelled volumetric flask and make up to the mark with mobile phase. Stopper and shake gently to mix.
c) Prepare a working solution by pipetting 10 mL of the mixed stock solution into

a labelled 100 mL volumetric flask and make up to the mark with mobile phase. Stopper and shake gently to mix.

The working solution contains 5 μg mL^{-1} of each of the components (2 μg mL^{-1} of uracil).

NB: It is possible to buy test mixtures to check column performance.

Perform the following experiments to determine the best operating conditions to achieve good separation and reliable results.

5.1.1 Flow Rate

Start the pump and allow the system to equilibrate, then inject the sample. Record the retention times and peak areas of the analyte peaks.

Then vary the flow rate by ±25% and see what effect this has on retention times, peak areas and peak separation (and back pressure).

Decide on which flow rate achieved good separation and short analysis time.

5.1.2 Mobile Phase Composition

Repeat the experiment but this time vary the mobile phase composition (*e.g.* 55% acetonitrile/45% phosphate buffer).

Decide on an optimum mobile phase composition for achieving good separation.

5.1.3 Detector Wavelength

Repeat the experiment but this time use a different detector wavelength (*e.g.* 254 nm ± 15 nm).

Tabulate the effect this has on the peak areas of the various components. Decide how critical it is to control wavelength (look at the spectrum in Figure 14 for a guide and if equipment is available record the UV spectrum of the components).

5.1.4 Repeatability

Perform five replicate injections using the operating conditions that give the best separation and a reasonable analysis time.

Calculate the mean and standard deviation of the retention times and peak areas for each component. Use these figures to estimate what the repeatability would be at the 95% confidence level (≈ 2 standard deviations) using this system.

6 Calibration

Calibration of the HPLC system is important to obtain accurate results. It is also recommended to run frequent checks to ensure that all instrumental parameters are not drifting. Standards and/or test mixes can be run as often as necessary as a part of Quality Control (QC). When components, such as the mobile phase, have been changed, calibrate the instrument immediately and ensure that the repeatability of results are within the specified tolerances, before it is to be used for analysis. You should keep all chromatograms from the standard runs and note changes in the operating parameters which have been made to overcome problems (if any). This not only allows you to frequently compare chromatograms (to identify any obvious problems) but also enables you to build up a profile on the history of the column. It is also advisable to plot the results from QC samples on a control chart, to monitor progress and to help identify any potential problems.

6.1 External Standard

External standards are analysed with test samples to quantify the analyte concentrations in the sample 'external standardisation'. The standards can be used to establish response factors for individual components within a sample. Generally a range of analyte concentrations is expected from the samples which require testing. Therefore, by preparing a series of calibration standards of known concentration and running these on the HPLC system, a calibration plot of detector response *versus* analyte concentration can be obtained. This series should at least cover 120% of the full range of concentrations of the samples to be tested and should also contain at least six calibration levels. From this plot, the concentration of unknown samples can be determined.

6.2 Internal Standard

Internal standardisation involves the addition of a known amount of a compound to the sample prior to its injection onto the column. Although not widely used in HPLC (because of the predominant use of auto-injectors) the advantages of internal standards include:

- any loss of sample in the preparatory stages will also be accounted for by a similar loss on the amount of internal standard
- the quantification procedure is less dependent on the injection technique, *e.g.* inconsistency in the amount of sample from manual injections

However, the problems with using internal standards include:

- finding a suitable compound which is readily available
- complications due to the presence of an additional peak in the chromatograph

7 Problem-solving

7.1 Introduction

One of the tasks you will undoubtedly face during HPLC operations is the necessity to troubleshoot. Many problems, when first encountered, appear to be baffling but once identified, they can usually be resolved by checking and performing remedial actions, *e.g.* fix a leaking connection.

7.1.1 Daily Checks and Fault Checking

If the problem has been identified by observations from the chromatogram:

- run the sample again and/or employ a standard to check if there is a fault
- check solvents are filtered and degassed
- check settings are correct for the method, *e.g.* flow rate, detector wavelength *etc.*
- check the system flow is correctly set
- observe pressure during analysis to check for abnormality

Diagnose the fault (if possible) and seek advice where necessary. If a hardware component is to be changed, carefully dismantle and throw away defective items, reassemble with replacement of the damaged parts and record changes.

7.1.2 Troubleshooting

The following figures show some of the typical problems encountered and the tables provide possible solutions. Further detailed information can be obtained from the references indicated for specific problems.

7.1.3 Unusual Peak Shapes

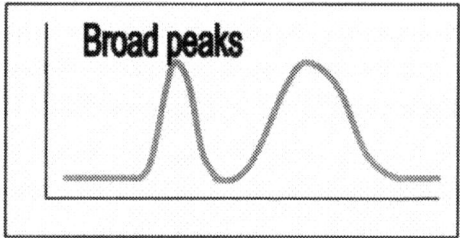

Figure 16 *Broad peaks*[11,12]

Table 5 *Broad peaks*

Cause	Cure
Sample is too concentrated	Dilute sample
Large injection volume	Reduce injection volume
Column deterioration	Wash column, replace column and use guard column
Dead space	Check fittings are secured properly

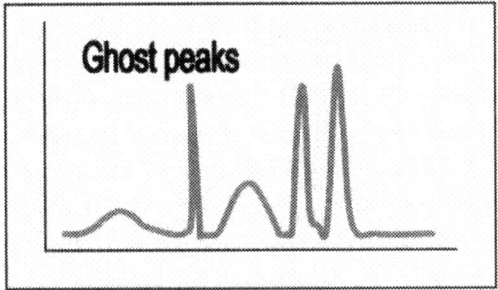

Figure 17 *Ghost peaks[13–15]*

Table 6 *Ghost peaks*

Cause	Cure
Contamination	Flush column with solvents to remove contaminant
Compound from earlier injections	Flush column with strong solvent
Unknown	Apply sample clean-up or check purity of mobile phase

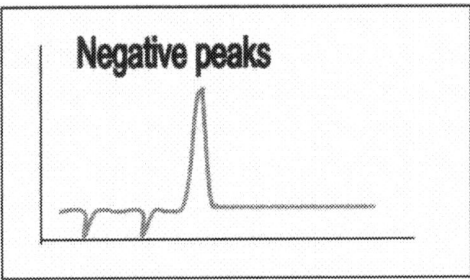

Figure 18 *Negative peaks*

Table 7 *Negative peaks*

Cause	Cure
Mobile phase absorbance is larger than sample absorbance	Use mobile phase that does not absorb at the wavelength used
Recorder connections	Check polarity of recorder connections

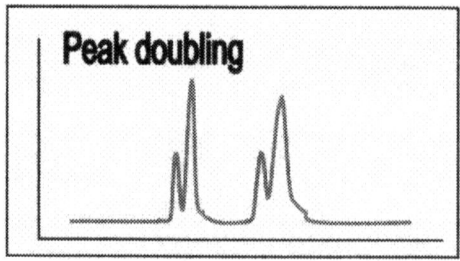

Figure 19 *Peak doubling[13–16]*

Table 8 *Peak doubling*

Cause	Cure
Co-elution of interfering compound	Improve sample clean-up or use pre-fractionation Adjust selectivity by changing mobile/stationary phase
Column overload	Use higher capacity stationary phase, increase column diameter or decrease sample load
Channelling in column	Replace column

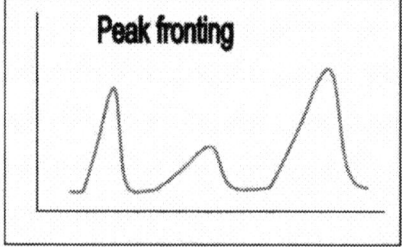
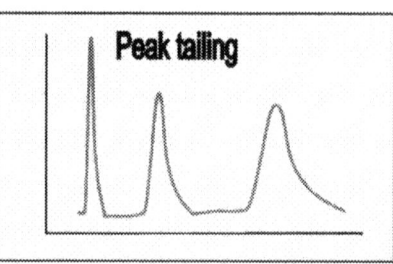

Figure 20 *Fronting and tailing peaks[15]*

Table 9 *Peak fronting and tailing*

Cause	Cure
Peak fronting	
Overloaded column	Decrease sample size, use higher capacity column
Peak tailing	
Overloaded column	(see peak fronting)
Basic compounds which cause silanol interactions	Use competing base such as triethylamine or base deactivated silica reversed phase column
Column deterioration	Use guard column/apply less vigorous conditions

7.1.4 Baseline Spikes

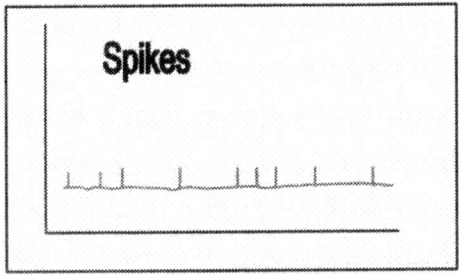

Figure 21 *Baseline spikes*

Table 10 *Baseline spikes*

Cause	Cure
Air bubbles in the mobile phase and/or detector	Degas mobile phase Ensure that all fittings are tight
Electrical interference	Identify and remove sources, *e.g.* thermostatted equipment Poor electrical configuration or bad connections
Column deterioration	Change column/use guard column

7.1.5 Clipped Peaks

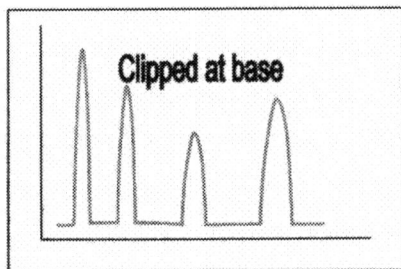 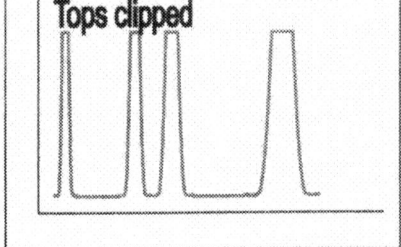

Figure 22 *Clipped peaks*

Table 11 *Clipped peaks*

Cause	Cure
Clipped at bottom	
Integrator zero set too low	Set zero correctly
Detector drifts below zero	Use auto-zero function
Clipped at top	
Detector range is too sensitive	Set less sensitive range
Column overload	Dilute sample

7.1.6 Inconsistent Retention Times

Table 12 *Fluctuating retention times[17]*

Cause	Cure
General	
Equilibration time of new mobile phase is insufficient	Pass 20 column volumes of new phase through the column
Selective evaporation of mobile phase component	Cover solvent reservoirs Use less vigorous helium degassing
Temperature fluctuations	Use thermostatted column oven
Decreasing retention times	
Increasing flow rate	Check and if necessary reset pump
Degradation of silica stationary phase	Ensure that the mobile phase pH is within the column's tolerance limit (typically between pH 2 and 8)
Temperature fluctuations	Use thermostatted column oven
Increasing retention times	
Decreasing flow rate	Check and if necessary, reset pump Check for leaks within the system
Temperature fluctuations	Use thermostatted column oven

7.1.7 *Drifting Baseline and Noise*[18]

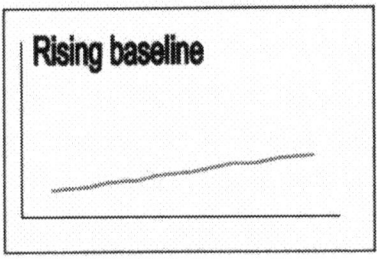

Figure 23 Baseline problems

Table 13 *Drifting baseline and noise*

Cause	Cure
Baseline drift	
Contamination build-up/column ageing	Flush column/use guard column Change column
Temperature changes	Use a thermostatted column
Noise	
Continuous – detector lamp problem or dirty flow cell	Replace UV lamp (should last for about 2000 hours) Clean and flush flow cell
Random – contamination	Flush column with strong solvent/clean-up sample

7.1.8 *Leaks and Pressure Changes*

Table 14 *Problems with leaks and fluctuating pressure*

Cause	Cure
General	
Column – loose fitting	Tighten or replace fitting Disassemble fitting, rinse and/or replace ferrules
Poor connection between components	Check fittings and connectors – clean, align or replace defective components
Fluctuating pressure	
Air in pump	Degas solvent
Leaking pump	Replace or clean check valves Replace pump seals

(*continued*)

Table 14 *continued*

Cause	Cure
Decreasing pressure	
Insufficient flow from pump	Loosen cap on mobile phase reservoir
Leak in lines from pump to solvent reservoir	Tighten or replace fittings
Leaking pump	Replace or clean check valves
	Replace pump seals
Increasing pressure	
Blocked flow lines	Systematically disconnect components from detector end to column end to find blockage
	Replace or clean any blocked component(s)
Particulate build up at head of column	Replace or clean frit
	Install 0.5 μm porosity in-line filter between pump and injector to eliminate mobile phase contaminants or between injector and column to eliminate sample contaminants
Water/organic solvent systems – buffer precipitation	Ensure mobile phase compatibility with buffer concentration
High back-pressure	
Column blocked with irreversibly adsorbed sample	Improve sample clean-up
	Use guard columns
Column particle size too small	Use larger particle size
Mobile phase viscosity too high	Use solvents with lower viscosity
Salt precipitation, *e.g.* in reversed phase systems with high concentration of organic solvent in mobile phase	Ensure mobile phase compatibility with buffer concentration
	Decrease ionic strength of the mobile phase
	Premix mobile phase

8 Data Handling

8.1 Recording, Manipulating and Reporting Data

Common mistakes are often made in the recording, manipulating and reporting of data. Computers, spreadsheets and laboratory information systems have revolutionised data handling procedures but have also introduced additional areas for potential mistakes.

8.1.1 Important Considerations

When using computer software for recording and analysing data, the analyst must know:

- how to use HPLC software, to programme and/or alter conditions for sample analysis
- how to integrate the chromatogram to obtain peak areas using appropriate data processing parameters (see Section 7 of the Gas Chromatography guide)
- how to perform a calibration, *e.g.* how to:
  - enter data into a calibration table
  - use the software to calculate the calibration curve and process sample data
  - archive sample chromatograms, calibrations *etc.*, to act as a back up
  - use, *e.g.* quality control samples, to check the calibration

The analyst should also be aware that different software packages may use different integration algorithms. Therefore, a method will need to be validated for each software package.

Check the calibration by calculating the results using, *e.g.* a calculator. Further tips on how to analyse data by using computer spreadsheets are described in reference 19.

8.2 Keeping Notebooks

It is important to keep a record of your experiments, observations and results, and a good way of doing this is by recording information in a book. You may wish to keep more than one book; one to record your individual work and another used as a system logbook to record information about the HPLC system. The analyst should:

- record data correctly and legibly
- ensure that data are correctly copied from one place to another without loss or transposition of results
- record data using the correct units, *e.g.* μg, mg or g for mass data
- include enough information to allow someone else to repeat the experiment.

Guidance on how to maintain records and keep logbooks is discussed in reference 20.

9 A Guide to Finding Information

The amount of information on HPLC that is available is overwhelming. Data are readily available and most sources are written by experts, who have extensive knowledge of the subject.

Books

HPLC and general analytical text books provide useful information on the theoretical principles. Some useful texts:

- V. Meyer, *Practical High Performance Chromatography*, 2nd Edition, Wiley, 1999, ISBN 0 03020 293 0.
- S. Lindsay, *High Performance Liquid Chromatography – ACOL Series*, 2nd Edition, Wiley, 1992, ISBN 0 47193 115 2.

HPLC Manufacturers and Suppliers

Consultation with experts from HPLC manufacturers and suppliers is a common way to seek advice. This approach is extremely useful for obtaining information on the range of products available and guidance on troubleshooting. Manufacturers' catalogues are often useful sources of information and many of these also provide good background details on theory and application.

Internet

The Internet allows access to information in an instant and there are many useful sites that provide helpful guides. Some useful sites include:

> http://hplc.chem.shu.edu
> http://kerouac.pharm.uky.edu/arsg/hplc/hplcmytry
> http://www.lcgcmag.com

Training Courses

Training courses allow you to learn through lectures, demonstrations and in many cases, practical exercises. Discussions of applications and problems with the trainer(s) and fellow trainees are additional ways to learn.

You can find out what is available by:
- contacting HPLC manufacturers
- reading HPLC literature such as *LCGC* magazine

---

10 References

1. V. Meyer, *Practical High Performance Liquid Chromatography*, Wiley, 1999, ISBN 0 47198 373 X.
2. C. F. Poole and S. Poole, *Chromatography Today*, Elsevier, 1991, ISBN 0 444 89161 7.
3. D. A. Skoog, D. M. West, F. J. Holler and S. R. Crouch, *Analytical Chemistry – An Introduction*, Saunders College Publishing, 1999, ISBN 0 03020 293 0.
4. S. Lindsay, *High Performance Liquid Chromatography – ACOL Series*, 2nd Edition, Wiley, 1992, ISBN 0 47193 115 2.
5. E. Prichard, G. Mackay and J. Points (eds), *Trace Analysis: A Structured Approach to Obtaining Reliable Results*, Royal Society of Chemistry, 1996, ISBN 0 85404 417 5.
6. J. W. Dolan, *LCGC International*, 1997, **10**, 3, 150–156.
7. J. W. Dolan, *LCGC International*, 1998, **11**, 5, 292–297.
8. R. G. Wolcott and J. W. Dolan, *LCGC International*, 1999, **12**, 5, 250–264.
9. R. E. Majors, *LCGC International*, 1999, **12**, 4, 212–221.
10. P. Bedson and D. Rudd, *Accred. Qual. Assur.*, 1999, **4**, 50–62.
11. J. W. Dolan, *LCGC International*, 1998, **11**, 4, 199–202.
12. N. Wilson, J. Kern and J. W. Dolan, *LCGC International*, 1998, **11**, 6, 350–354.
13. R. D. McDowell, *LCGC International*, 1997, **10**, 6, 358–359.

14 Y. Egi and A. Ueyanagi, *LCGC International*, 1998, **11**, 3, 142–147.

15 J. W. Dolan, *LCGC International*, 1997, **10**, 9, 570–572.

16 J. H. Huang, T. Culley and J. W. Dolan, *LCGC International*, 1999, **12**, 4, 208–221.

17 J. W. Dolan, *LCGC International*, 1997, **10**, 10, 640–646.

18 K. L. Christianson and J. W. Dolan, *LCGC International*, 1997, **10**, 11, 714–718.

19 G. I. Ouchi, *LCGC International*, 1997, **10**, 7, 427–430.

20 J. W. Dolan, *LCGC International*, 1997, **10**, 4, 216–220.

Lightning Source UK Ltd.
Milton Keynes UK
UKHW020627010421
381362UK00004B/67